HEINKEL He 280

The World's First Jet Aircraft

Edited by Joachim Dressel/Manfred Griehl/Jochen Menke

SCHIFFER MILITARY HISTORY
West Chester, PA

Photo Credits
Bekker Collection
Creek Collection
Dabrowski Collection
Emmerling Collection
Francella Collection
Griehl Collection
Heinkel Collection
Lutz Jr. Collection
Menke Collection
Mohr Collection
Nowarra Collection
Selinger Collection
Schliephake Collection
Forschungsgruppe Luftfahrtgeschichte e.V. Collection
EHAG Factory Photos

SOURCES:
DFS Correspondence
Ernst Heinkel AG Files
LwFüst Correspondence
RLM Correspondence

Translated from the German by Dr. Ed Force.

Library of Congress Catalog Number: 91-66338.

Printed in the United States of America.
ISBN: 0-88740-344-1

This title was originally published under the title,
Heinkel He 280 Band 108,
by Podzun-Pallas Verlag, Friedberg.

Published by Schiffer Publishing, Ltd.
1469 Morstein Road
West Chester, Pennsylvania 19380
Please write for a free catalog.
This book may be purchased from the publisher.
Please include $2.00 postage.
Try your bookstore first.

This factory airfield gives an idea of the numbers of aircraft types that came into being under the direction of Ernst Heinkel.

Foreword

Some 75 years ago, in July of 1911, the 23-year-old Ernst Heinkel took off near Stuttgart on the successful first flight of a biplane that he had designed and built. In 1911 he joined Albatros as a designer, and by 1914 he was already the chief designer of the Brandenburgische Flugzeugwerke. On December 1, 1922 Heinkel founded his own firm in Rostock-Warnemünde. As a man with an almost unique sense of the new, he developed, after having made a name for himself with the He 70 and He 111, the He 176, forerunner of the later jet and rocket fighters. Then came the He 178 with the He S3B to power it, developed by powerplant specialist Dr. von Ohain. The goal-oriented evaluation of all previous results at the end of the 1930s then pointed the way to the first twin-jet plane, the He 280.

Along with the use of two He S8A engines, the first-time use of high-speed profiles, the new nose wheel scarcely noticed in Germany, and a compressed-air catapult as a life-saving device for the pilot, the plane as a whole was new.

This and next page: The world's first jet plane, the He 178, as a model. Note the filigree design of the landing gear.

Only the too-long retention of double rudders and the very unfortunate course of powerplant development for all the major manufacturers prevented this superior design from being put into series production. There remained a few test models, which were used in high-speed flight experiments until the last days of World War II and provided the DFS with valuable data for a series of interesting project studies.

Development

On May 15, 1936 the talented physicist Dr. Hans-Joachim Pabst von Ohain joined the Heinkel Flugzeugwerke to realize his plans for a high-performance turbojet engine. A demonstration motor would already be finished in 1935; it led to the He S3A, which was already producing 450 kph in the spring of 1938.

On August 27, 1939 factory pilot Erich Warsitz took off at Marienehe with the He 178, which was powered by an He S6 turbine. After its introduction on November 1, Udet and his staff were very excited. The way was cleared for the twin-jet pursuit fighter with turbojet (TL) engines. Following the guidelines issued by the Reich Air Ministry (RLM) on January 4, 1939, the Heinkel factory went to work at once. The first project studies anticipated a plane with a wing surface of 21.5 square meters, a flying weight of 4.25 tons and a speed of Mach 0.82. Three MG 151/20 guns were planned for.

Three of the few original photos of the He 178. The plane made its first flight on August 27, 1939, under pilot Erich Warsitz.

Above: The He 280 V1 (DL+AS) coming in to land. It took off under tow by an He 111. The powerplant dummies were necessary to achieve realistic flying behavior.

The top leaders of the German Reich tensely watch the takeoff of this new propellerless airplane, in some cases skeptically: from left to right, Keitel, Göring, Hitler.

Under the leadership of Messrs. Regner and Mach, two different mockups were built under the strictest secrecy in June of 1939. In addition, Heinkel worked on a weapon mockup with up to four automatic weapons. As of July 1939 studies for a plane with a nose wheel were made. Two months later, on September 26, 1939, representatives of the RLM, the Rechlin Test Center and the Luftwaffe General Staff examined the mockup of the pursuit fighter plane, now called He 280 instead of He 180. The entire cockpit and the arrangement of the weapons convinced the Luftwaffe. A mockup of the radio system (FT) was to be set up by October 1, 1939.

During the development discussions of October 20, 1939, at which Dr. Heinkel, directors Lusser and Pfistermeister, and Ernst Udet were present, the latter said that he was a fanatic believer in the success of Heinkel's work. The firm should build its jet plane alone, unlike Messerschmitt, without an outside supplier of powerplants! But the turbojet engines caused production problems for Heinkel as they did for all the German manufacturers of them. Thus Heinkel could not plan on model testing of the He S8A engine until August of the following year, since the rear turbine bearing was not yet sufficiently developed.

Until the engines had gained the necessary operating safety, other means of testing were utilized at various times, so as to eliminate delays as much as possible. For example, the firing tests could be concluded successfully at Marienehe by using a functioning mockup of the nose weapons with three MG 151/20 guns. Testing of the compressed-air ejection seat began as of June 15, 1940. After tests with sandbags ejected from the meanwhile modified He 280 mockup had caused no problems, tests with Heinkel staff members were begun in June. The first test person was Mr. Ohmann, followed by 75 others. Among them was a woman, the 30th test person. With a compressed-air pressure of 80 to 120 atu, the test persons were ejected 6 to 7 meters out of the cockpit.

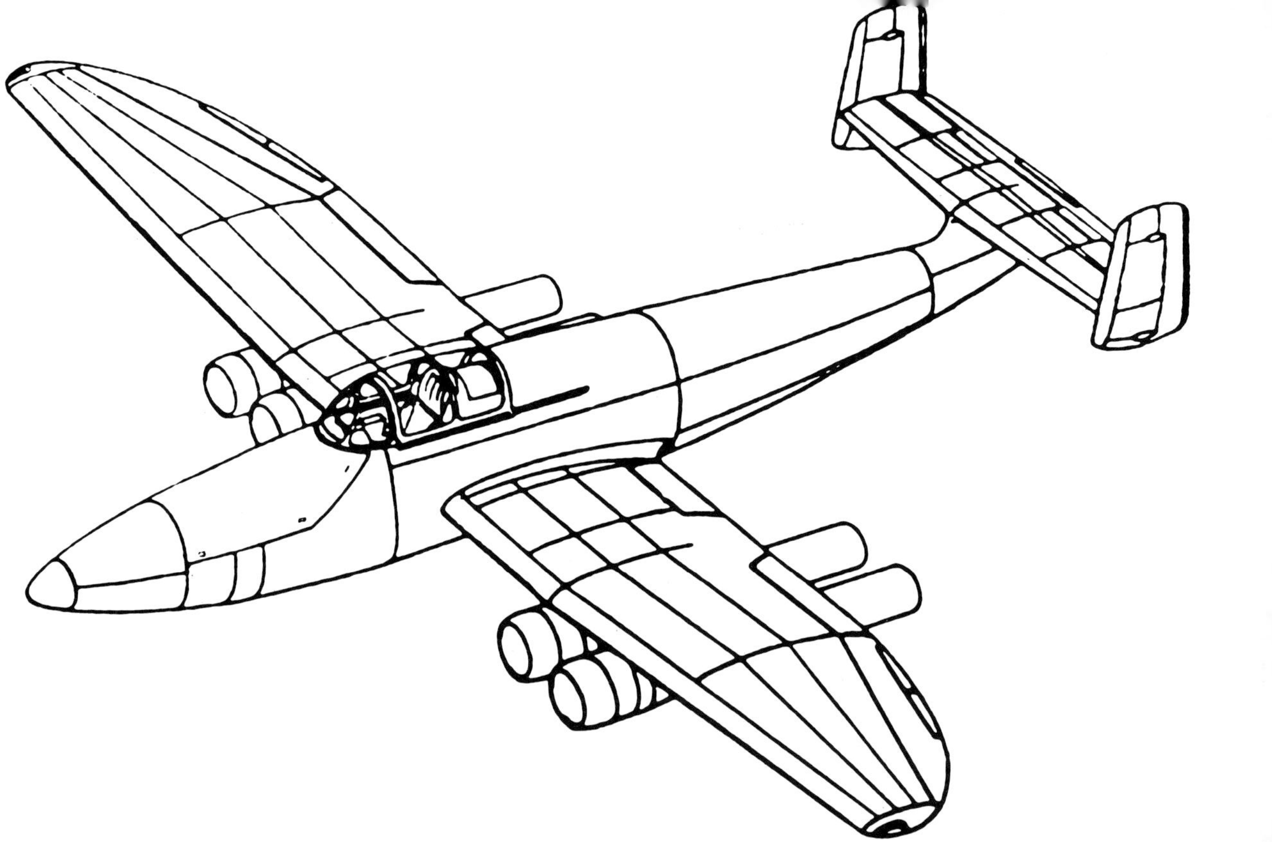

Above: The He 280 V1's original powerplant layout was to look like this.

Right: A wind-tunnel model used for measuring wing vibrations.

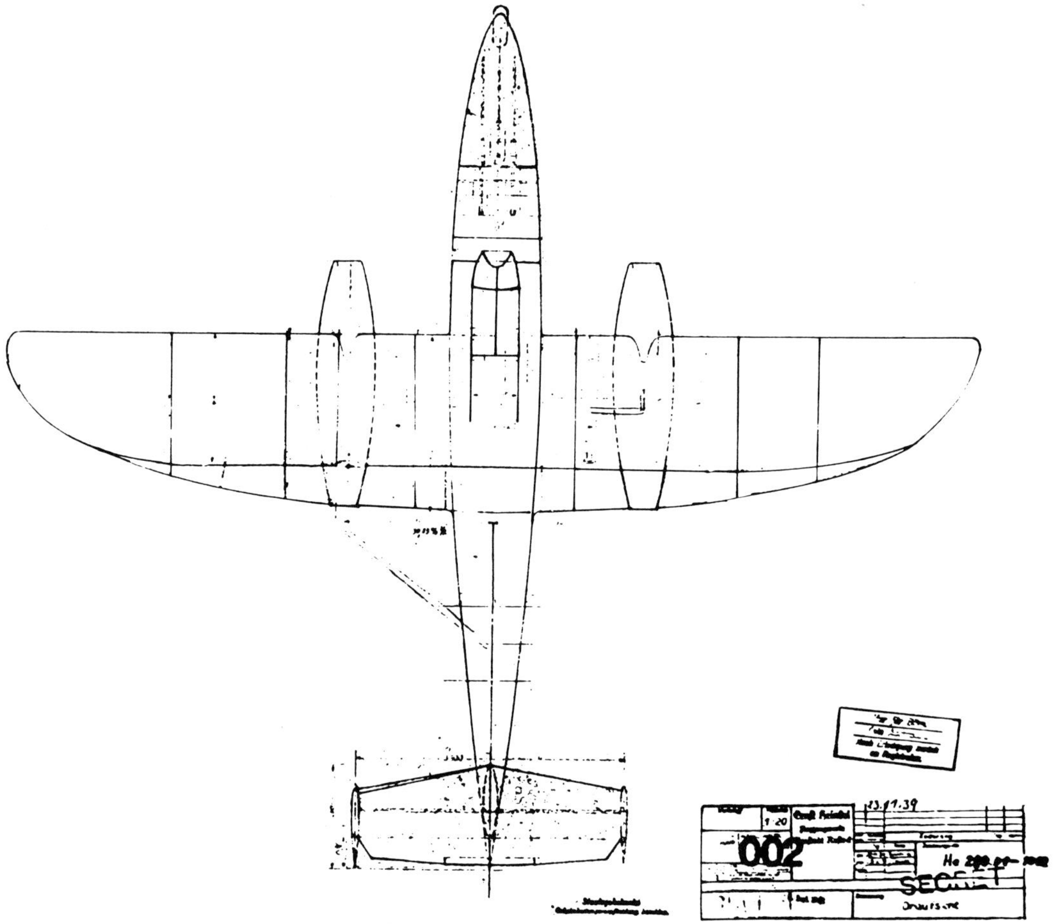

Above: The He 280 V1 in flight.

Left: Drawing of the first He 280 design of November 23, 1939.

Lower left: Front view of the first design.

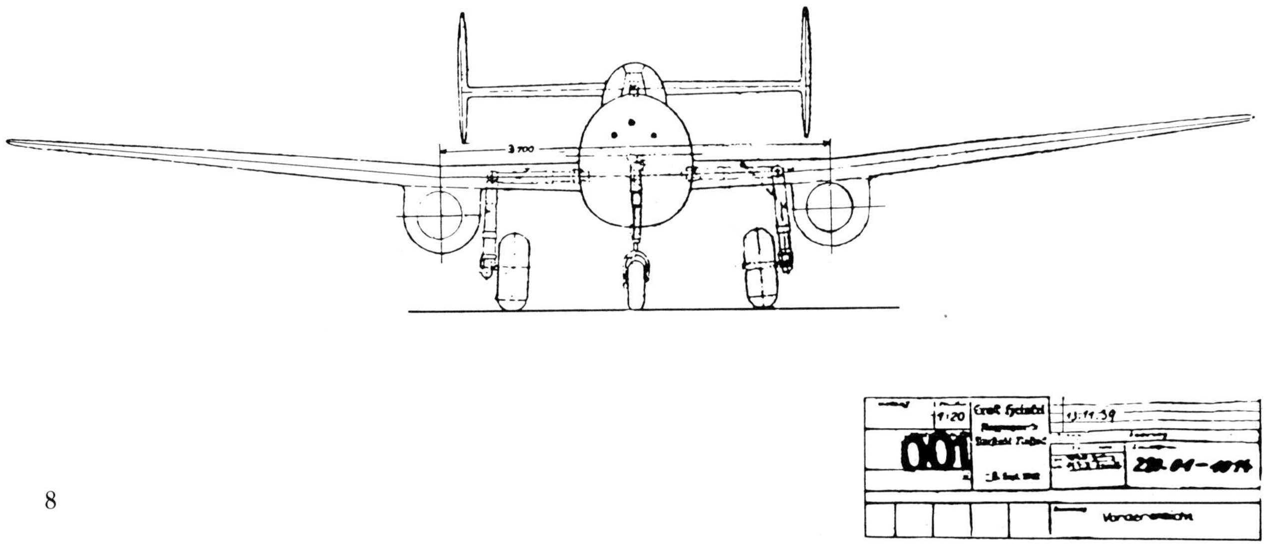

The first rolling tests of the He 280 prototype could begin at Rostock-Marienehe on August 28, 1940. The actual flight testing then began on September 22, 1940 at Rechlin-Roggentin, when the He 280 V1, with master aircraft builder Bader at the controls, was towed by an He 111 B piloted by Deutschmann for its first powerless gliding testing. Altitudes between 2500 and 4000 meters were generally attained, giving the He 280 B1 the chance to carry out actual flight testing down to about 1000 meters. As of October 10, 1940 the plane was back at the Heinkel works in Marienehe for further testing, and flights were made until at least May 22, 1941, to test not only the center of gravity and polar orientation but also the landing gear and flaps, as well as for stunt flying. After the towing cable released prematurely on October 10, 1940, builder Bader had to make a belly-landing, which caused a brief interruption of testing before the He 280 V1 — without the least tendency to flutter — attained speeds of 620 to 800 kph.

At the end of 1940 the lacking engines were still the chief subject of development discussions: The radial compressor and the fuel injection jets had to be changed several times, but the intended thrust of 700 kp was not yet attainable; only 550 kp could be attained at 13,000 rpm. At first no one expected the first flight of a powered He 280 before April of the following year. The powerplant department planned to begin the first ten standing tests of the experimental He S8 V3 and V4 engines in mid-December of 1940.

During the discussion on December 9, 1940, Ernst Udet said of the He 280: "The plane itself is ready to fly, only the engines are lacking, but I have complete faith in its success!"

But skepticism spread, since the flight date of the first prototype with Heinkel engines had to be postponed again and again.

On December 12, 1940 the state of development was as follows: The He 280 V1 had made sixteen towed flights and reached a speed of 600 kph. The second test model was in its last stages of assembly, while preparation of components and individual parts for He 280 V3 to V5 was in full swing. The crash cells had also been finished, and testing of them was to begin in January of 1941.

After evaluating reconnaissance photos and pilots' statements, the Americans drew this picture of the He 280. The drawing's authenticity to the original is remarkable.

Above: The machine-gun mounts during weapons adjustment.

Upper left: Weapons testing with the front section of the He 280's fuselage.

Left: A close-up view of the weapons layout, seen from the left. The flag shows the weapons are loaded and ready to fire.

In this top view, the three machine guns and the ammunition supply can be seen.

Front view of the nose with three MG 151/20 ports.

He 280 V2, still without protective paint. The sheet-metal joints can be seen clearly. The different shades are caused by different types of metal.

To deal with the powerplant problem, Dr. Heinkel proceeded to purchase the Hirth Motorenwerke as a developmental and production factory for the new jet engines, a move that was ultimately approved by the RLM. Meanwhile, flight testing of the He 280 V1 went on steadily and attained speeds of 700 kph. By early February 1942 Heinkel had finished the crash tests with the fuselage cells and began static checking of the wings, control surfaces and landing gear. In addition to wind-tunnel tests at the German Test Institute for Aeronautics, in which Heinkel staff members took part, the factory began to prepare for the first flight of He 280 V2 with He S8A jet engines and to build the zero series of the He 280 high-speed pursuit fighter.

On March 30, 1941 the day had finally come: Fritz Schäfer taxied and took off in the first jet-propelled He 280 V2. Although the engine covers were removed and the landing gear and flaps remained deployed, the first flight went off without any problems. Schäfer only reached 250 kph with He 280 V2 that first day, but the speed increased in subsequent flights. Difficulties included the lack of an rpm indicator and direct thrust-jet adjustment. At first it was only possible to regulate the two engines' fuel intake by using the gas pedal. Until the urgently needed controls were developed, the pilots had to make do as best they could. Yet several transfer flights went off well, including one of He 280 V2 from Marienehe to the test center at Rechlin at 3000 meters elevation, with an average speed of some 650 kph. And this took place at a time when the first flight of the Me 262 (April 18, 1941) was only possible with the help of a Jumo 210G piston engine.

To put more He 280's in the air under

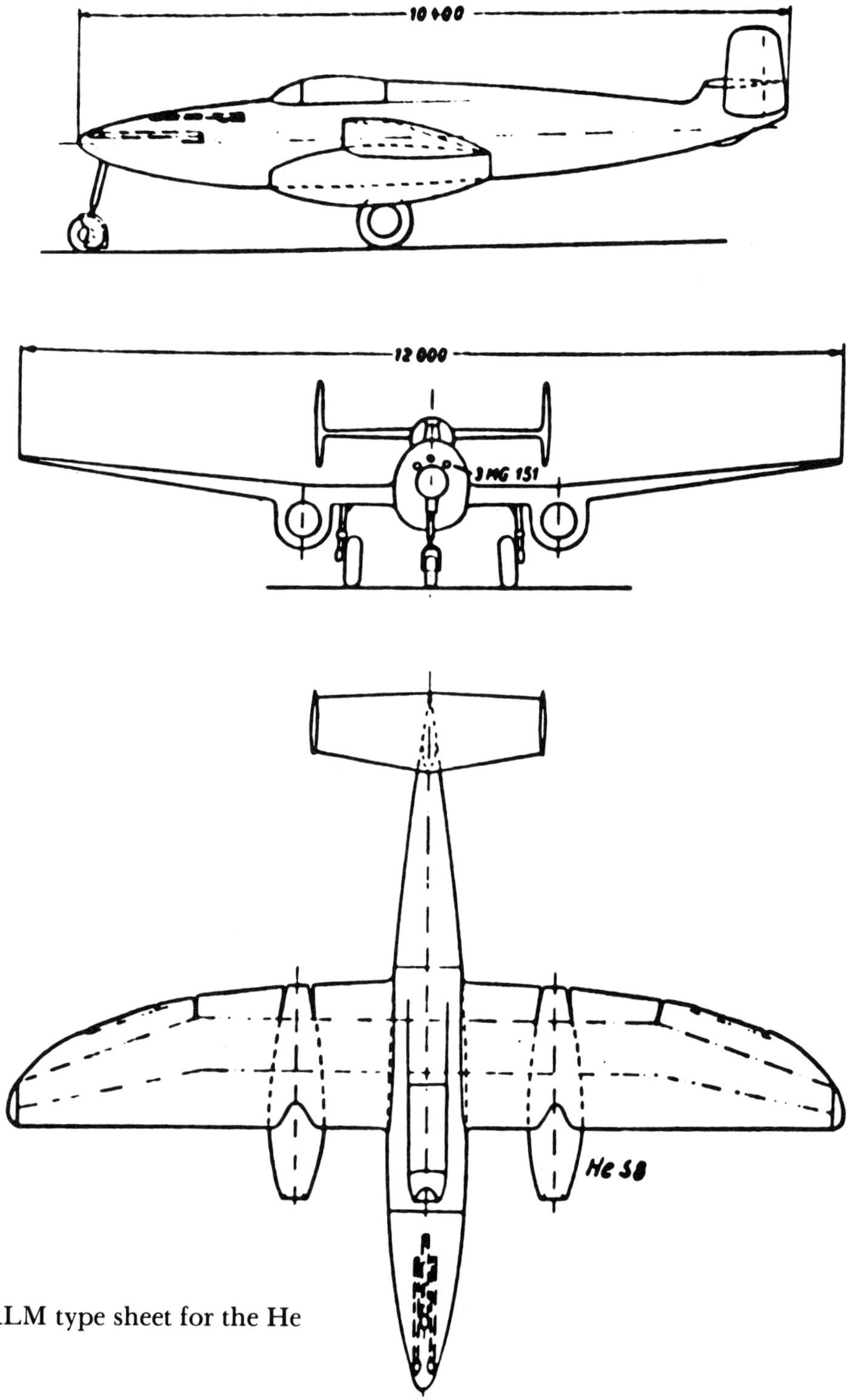

Right: The RLM type sheet for the He 280.

Above: He 280 V2 takes off, still without engine cowlings. This was done to avoid the danger of fire from fuel dripping onto the cowlings.

Below: He 280 V2 with Ernst Udet in the cockpit.

their own power in the shortest possible time, an interim solution was the use of Walter Pulso engines, two or three of which could be installed under each wing.

On July 11, 1941 Udet again inquired about the state of the He 280's development, for he wanted to have it available as an interceptor in the near future. He described its first flight as "very impressive." Unfortunately, the question of the He S8A engine's readiness for front-line service or for mass production could not be answered definitely.

Miracles cannot be made from one day to the next! The thorough development program could be carried out only step by step because of the new engine technology. The following modifications were planned:

— Aerodynamic improvements to the cell (glider),
— Engine development and testing,
— Flying characteristics and performance improvement,
— High-speed testing.

During the discussion between the Heinkel firm and the Luftwaffe General Staff, Majors Storp and Ditfurth reported:

"As war continues to develop at the present time, we cannot match the capacity of America and Britain quantitatively . . . so quality must be demanded at the cost of quantity. In order to be successful in the home war zone, good high-speed performance is particularly important."

Early in 1942 the RLM expressly demanded: "The He 280 must fly!" Two weeks later, on January 23, 1942, Ernst Heinkel called his directors and Messrs. Günter and von Ohain together. The RLM neither could nor would wait longer. Cell testing was to take place immediately, using several Argus jet engines lengthened by two meters and producing 150 kp of thrust. This simple powerplant had already been tested in the Go 145, Bf 109, Bf 110 and Do 217.

The Jumo 004-TL suggested by the RLM was likewise not immediately available, at least not before the next spring. On March 3, 1942 representatives of the Heinkel works were shown two of the Pulso engines running on test benches at the Argus factory in Berlin-Reinikendorf. Since the He 280 would usually fly faster than 180 kph, cooling problems were scarcely to be expected. For the take-off of an He 280, 2x4 Pulso jets would be necessary. By retaining the take-off on tow by an He 111, though, a total of four As 014 units would suffice for flight testing. At an altitude of three kilometers, an average speed of over 700 kph was estimated as of May 1942. In addition, a second test plane was to be fitted with eight Argus units. Besides this, the RLM requested accelerated installation testing for the new Jumo turbines, as well as the BMW P3302 jet engines. The latter, according to the BMW works in Munich, did not yet produce enough power to be installed in the He 280. Thus functional jet engines were available, but their power output was still much too low. The use of the Argus jets thus took on a higher and higher priority. In the conference held on June 1, 1942 the most urgent detailed questions could already be answered. In addition, the Argus factory supplied the first six jet units and began to devise a second set of four Pulso units. Preparations for the production of, at first, 300 of them were begun.

At the same time, though, Heinkel's own

He 280 V2 with Udet, Bader, Franke, Reitenbach, Lusser and Autz.

Ernst Udet with a camera. Wrapped up in flying as he was, he wanted to capture this flight on film.

He 280 V2 in flight with engine mockups. Remarkably, the He 280 never flew with camouflage paint.

jet engines were to be made ready for series production as fast as possible and, using an He 111 H, the effect of altitude on thrust, the engine controls and starting in flight were to be tested.

At first the test program called for the maiden flight of the He 280 V3 on July 5, 1942 to be made using two He S8A jet engines, later to be replaced by two of the more powerful Jumo 004 powerplants and weapons in the nose.

Shortly afterward, the He 280 lost its previous SS priority rating in the armament program.

On August 17, 1942, just as Me 262 V3 was taking off for its first flight at the Messerschmitt works, two conferences important for the further development of the He 280 were taking place: At one it was decided that He 280 V1 soon had to be tested at Rechlin with four Pulso jets, and that V2 and V4 had to be fitted with the more powerful Junkers jet engines. While the second He 280 was simply to receive the two Jumo 004 engines, Heinkel got the idea of fitting He 280 V4 with larger fuel tanks and an upgraded weapon system. This was to consist of either three MG 151/20 and two MG 131 guns or two MG 151/20 and two MG 131 plus one MK 108. To get results quickly, Ernst Heinkel personally ordered that the crash-test wings be repaired immediately for use on He 280 V4.

To give the RLM an overview of the future He 280 B's potential, the Heinkel works put together a data sheet in September of 1942, anticipating the use of He S8A, S8B and Argus jet units. A suggestion resulted in the simultaneous use of two He S8A's and two As 014 Pulso units. It was hoped that speeds of more than 860 kph could thus be attained. On the other hand, the He 280 B with He S8B engines and the new jet-engine grid controls would only reach 800 kph. With eight Pulso units, it was thought that a top speed of only 710 kph was possible.

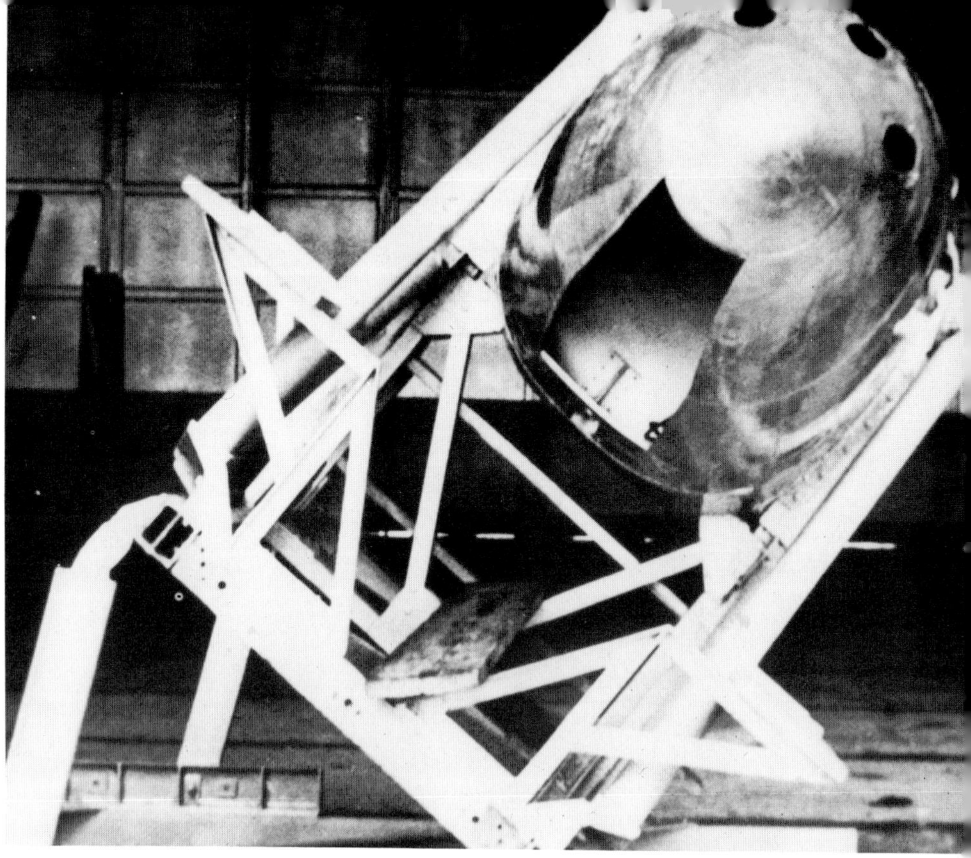

With this rack, adjustable in all directions, the functioning of the weapon system in any aircraft could be tested.

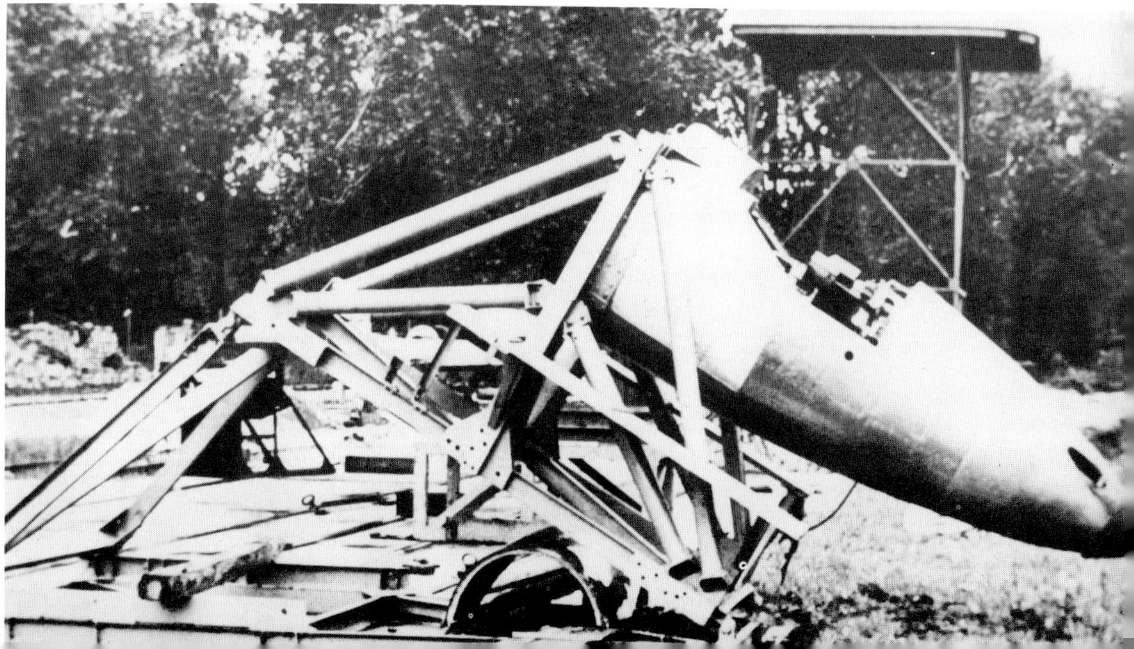

On account of engine trouble, He 280 V3 had to crash-land on February 8, 1943.

Final assembly of He 280 V3.

Factory testing of the He 162 at Schwechat.

Front view of He 280 V3; a good chance to compare cowled and exposed engines.

When Me 262 V2, powered by two Jumo 004 engines, took off for the first time on October 1, 1942, the last details involved in fitting He 280 V2 with the heavy Jumo turbines were being considered at Heinkel. For the future He 280 production version with Jumo 004 engines, though, there was not enough ground clearance, and the increased weight was also a disadvantage. Help could be gained only by strengthening and lengthening the landing gear, wings and fuselage. But there was no time to do this to He 280 V2. He 280 V1, powered by four Argus units, was to be finished by October 25, 1942, so that the plane could be demonstrated by the experienced Heinkel pilots Schuck and Schäfer at Rechlin. These plans were not to be realized.

The installation of BMW jet engines was planned to begin only with He 280 V8. The engines, though, could not be expected before the summer of 1943. The Schwechat airfield outside Vienna was to be the test center for the He 280, and the rest of the 20 ordered test planes were to be built there soon. All of them from He 280 V11 on were

The He S8A engine of He 280 V3. The intake has been covered with a wooden plate to keep dust out. This was removed before the engines were started.

to have the required larger fuel capacity.

Plans were made to equip 24 prototypes with Heinkel, Junkers and BMW jet engines, but it was determined on November 18, 1942 that for the time being only six test models with engines and one auxiliary unpowered craft for high-speed testing were to be built.

The Heinkel works planned the following at the beginning of 1943:

— He 280 V1: Begin testing with four Argus As 014 units,
— He 280 V2: Finish rebuilding for Jumo 004 engines at Schwechat,
— He 280 V3: Continue testing with He S8A engines,
— He 280 V4: Fitting with a pressurized cockpit and He S8A units,
— He 280 V5: Fitting with He S8 engines by January 15, 1943,
— He 280 V6: Rebuilding for BMW P 3302 engines planned,
— He 280 V7: Preparation for high-speed testing,
— He 280 V8: Fitting with Jumo 004 engines,
— He 280 V9: Fitting with BMW 003 engines.

But not all of these planes could actually be finished, though the individual components, with few exceptions, were already finished.

Second outside landing of He 280 V3 in March of 1943.

Front view of the towplane, with a good view of the nose wheel and towing fork, which was also used for steering.

V3 is towed to the airfield by a tractor. Maintenance personnel accompany the plane. The flight of a jet plane was still unusual at that time.

Above: He 280 V3 after its successful first flight.

Right: V3 taking off. The jet plane's high take-off speed required much practice for pilots accustomed to propeller planes.

Above: He 280 V3 during test flying.

Upper left: He 280 V3 again. Its paint looks darker in contrast to the white snow than in the previous picture showing it on the gray runway. Note the exit opening of the He S8A engine. The cable leads to the plane's outboard connection.

Lower left: Fritz Schäfer in the cockpit of He 280 V3.

Meanwhile the competition between the He 280 and the Me 262 had taken on a new dimension, to the advantage of the Messerschmitt fighter. The RLM was convinced that whichever manufacturer was first to install Jumo or BMW engines would be the winner. Up to then, Heinkel had wanted to improve the He 280 in terms of design and performance: Along with installing six machine cannons in the bow, bigger fuel tanks, a central rudder and a weapon system that could be dropped were to be included in the future series.

On December 11, 1943 Beauvais first flew the third He 280. Afterward he spoke very positively of the plane. So the nose-wheel performance was judged to be good, and the cell's load-carrying ability was also approved. Beauvais' opinions of the insufficient stability of the high axis and the wing vibration at speeds around 600 kph were not so good. The Me 262 was also said to experience problems at that speed. Along with too-high control forces and too-low stability around the high axis, the pilots at the Rechlin test center were also less than happy about the plane's taxiing speed.

Upper right: He 280 V3 in front of an He 177 with its cabin hooked on and its powerplants removed.

Right: A fuselage section with the attachment for installing an ejection seat. This piece of He 280 V4 was photographed at Vienna-Schwechat.

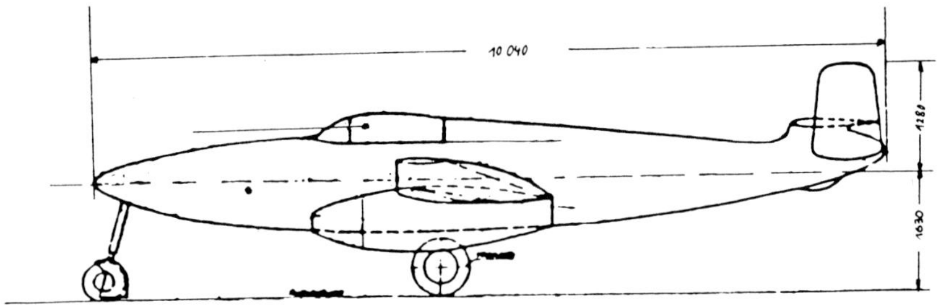

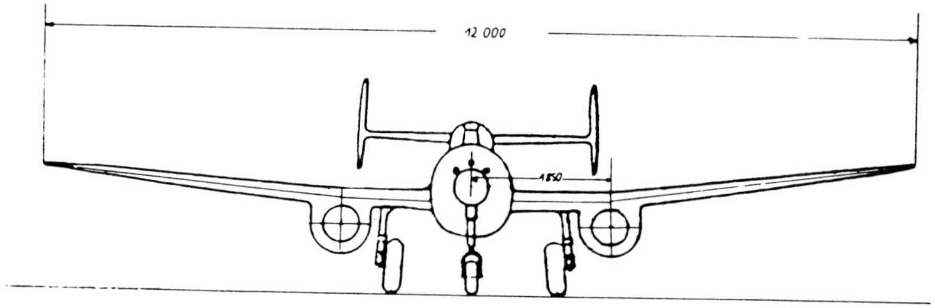

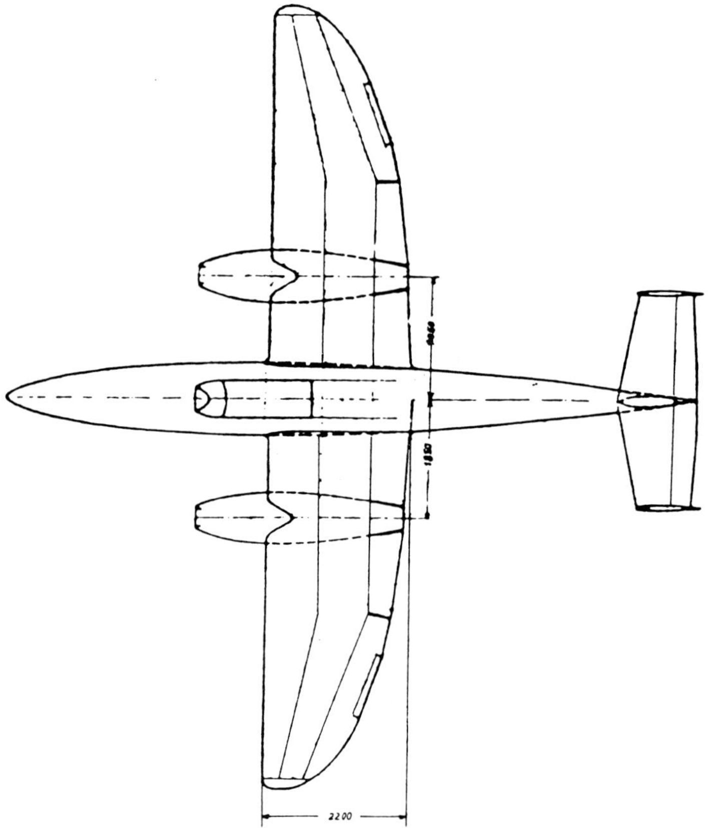

Heinkel factory drawing of September 8, 1942.

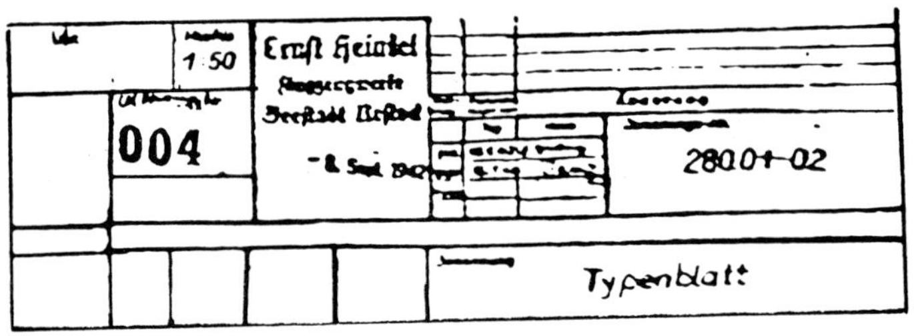

A considerable improvement in flying characteristics was expected from changing as quickly as possible to a central rudder, as was now urged by Dr. Heinkel. Shortly after that, Fritz Schäfer flew He 280 V2 for the first time. While being ferried to Rechlin, the plane reached a sustained speed of 650 kph at 2000 meters. On December 20, 1942 Ernst Heinkel wrote an optimistic letter to Milch: "I have believed in the He 280 for years and would like to undertake a few more improvements concerning the armament and the control surfaces!"

In December 1942, a brief description of the He 280 was issued to inform the Reich Air Ministry. Six test planes were described, of which two each were to be fitted with He S8A, Jumo and BMW engines. A version of the He 280 with the BMW P 3302 took in a particularly broad scope. Based on a thrust of 1360 kp in all, a speed of 815 kph was to be attained at 6000 meters. At this altitude it was hoped that a climbing speed of 10.7 meters per second could be attained. The calculated service ceiling was stated as 13,000 meters.

At the beginning of January 1943 the RLM, during a development conference, dealt with a letter from Heinkel saying that the He 280 with BMW P 3302 jet engines would be ready for series production not in two years, but in six months, if the necessary jet engines were made available. A zero series of 100 planes could then be started immediately. Production of the machines, as well as testing the B-series test models and any other test flying, were to be done by Siebel.

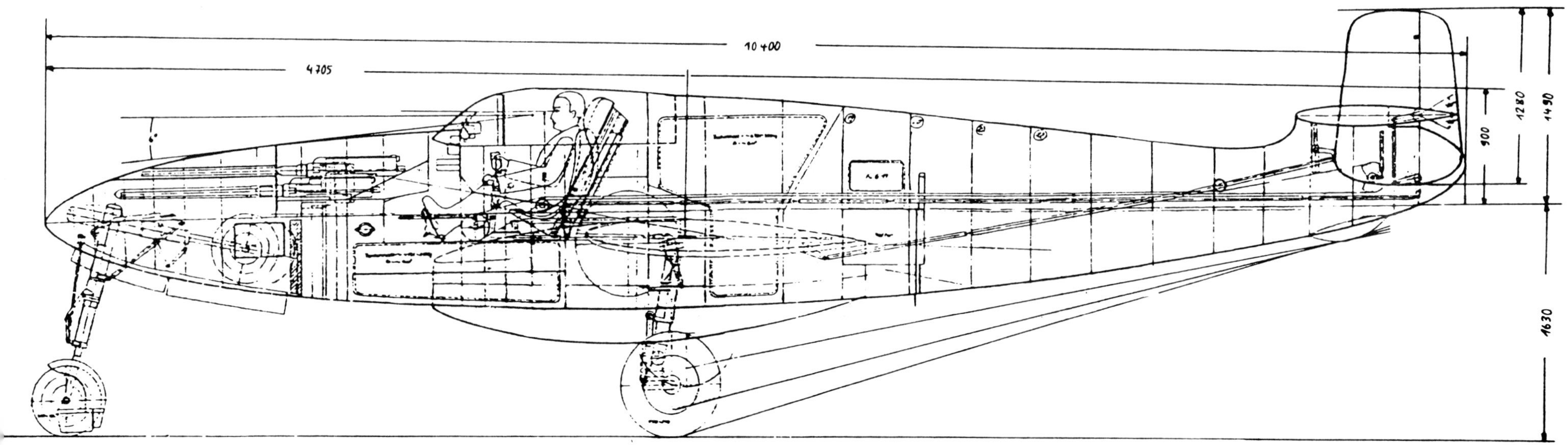

Side-view drawing of the He 280 with exact measurements.

On January 27, 1943 the He 280 was included in the Vulcan program, guaranteeing preferential assignments of workers and materials. A few days before that, the He 280 test program had suffered a major setback with the crash of the first test model. The plane was to be moved from Rechlin to Lärz to start flight testing with four As 014 engines. The take-off in heavy snow was difficult. The nose wheel could not be retracted and the controls broke down. Then there were problems with the towing-cable release, and the Argus factory pilot decided to leave via the ejection seat. This was the first emergency rescue achieved by using the ejection seat. The He 280 V1 made two

Three-way views of He 280 V7 with large jet duct. The plane underwent an extended flight testing, with a total of 115 towed flights.

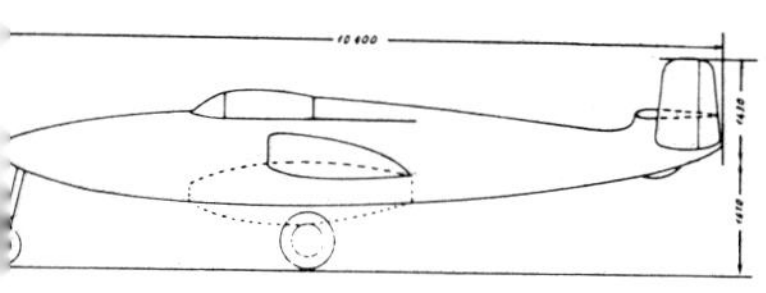

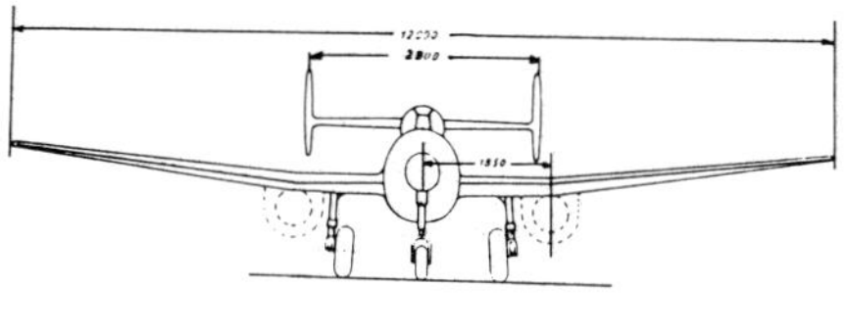

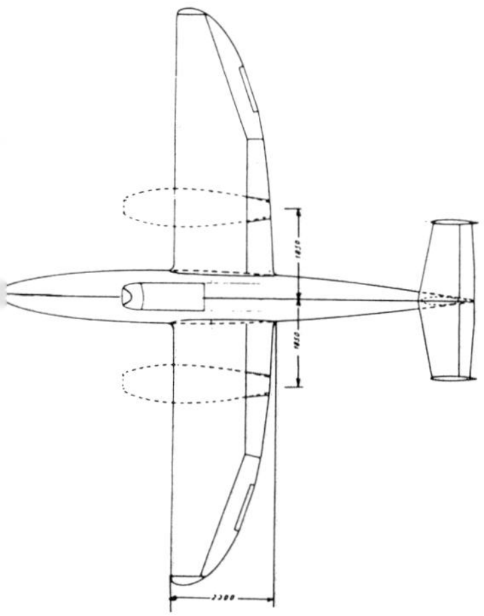

Flügelfläche *F = 21,45 m²*
Spannweite *b = 12,00 m*
Flügelstreckung *Λ = 6,82*

Wing surface: F = 21.45 sq.m.
Wingspan: b = 12.00 m
Wing aspect ratio: A = 6.82

He 280 V7 (ohne Triebwerke)

He 280 V7 (without engines)

Above: Drawings of He 280 V7.

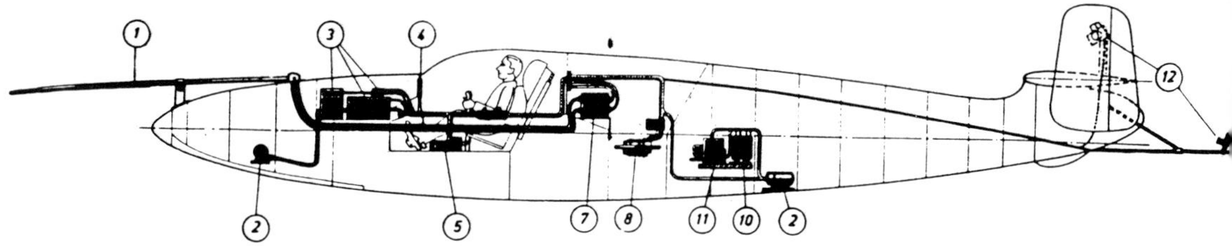

Einbauten in He 280 V7 für Abkippmessungen

Components installed in He 280 V7 for pitch measurement

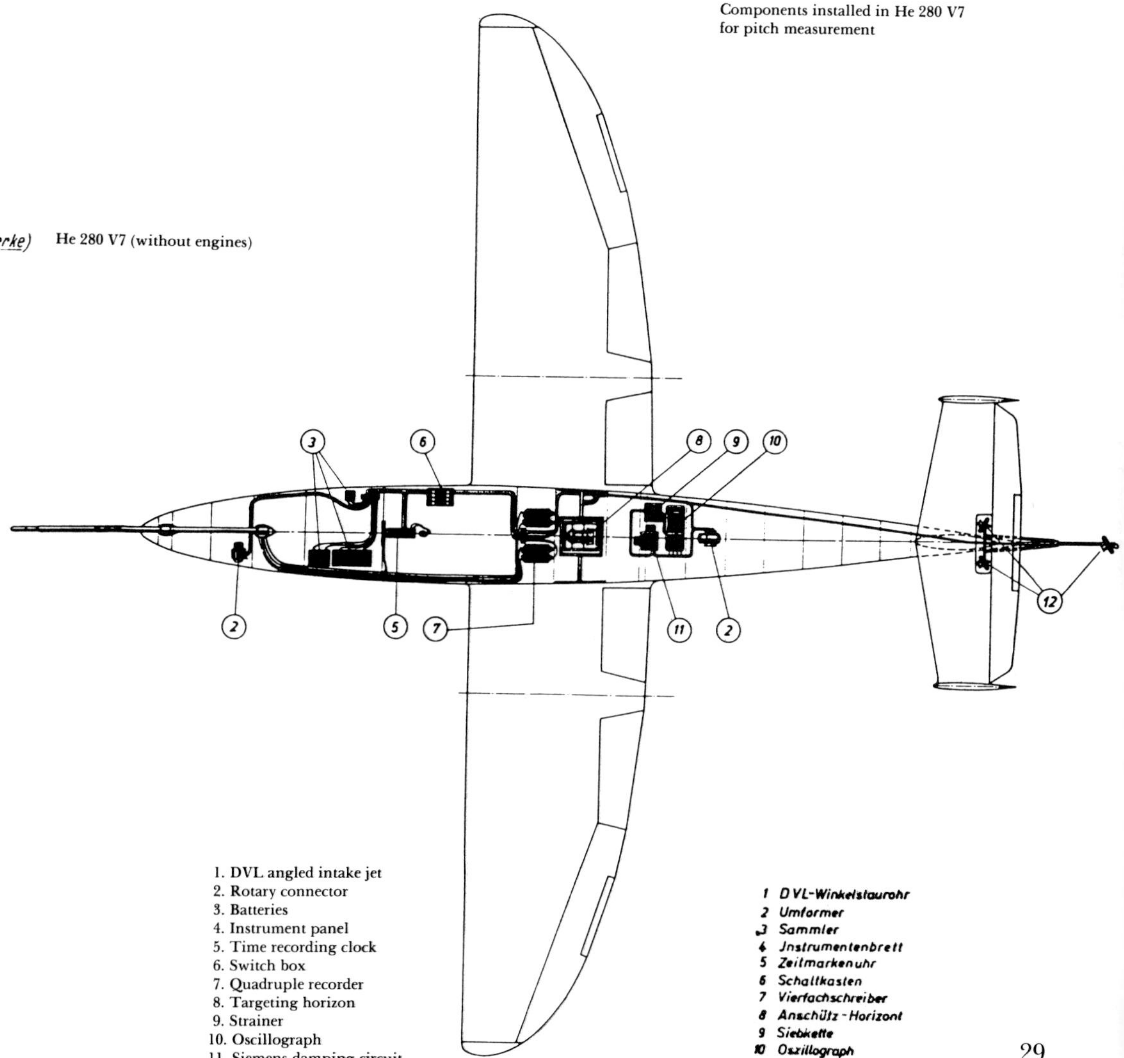

1. DVL angled intake jet
2. Rotary connector
3. Batteries
4. Instrument panel
5. Time recording clock
6. Switch box
7. Quadruple recorder
8. Targeting horizon
9. Strainer
10. Oscillograph
11. Siemens damping circuit
12. Robot cameras

1 DVL-Winkelstaurohr
2 Umformer
3 Sammler
4 Instrumentenbrett
5 Zeitmarkenuhr
6 Schaltkasten
7 Vierfachschreiber
8 Anschütz-Horizont
9 Siebkette
10 Oszillograph
11 Siemens Dämpfungskreisel
12 Robot-Kameras

Right: Drawings of the components installed in He 280 V7 for pitch measurements. Cameras were mounted on the rear bar.

loops before it landed and was almost completely destroyed. An experienced pilot, according to the investigation report, surely could have saved the test model. Scarcely a month later, on February 8, after ten successful test flights had been made, a long jet of flame shot out of the right He S8A engine and its power decreased abruptly. Fritz Schäfer closed the fire valve and crash-landed, and He 280 V3 was only slightly damaged. By May 21, 1943 the plane was flying again. Meanwhile He 280 V2 was being remodeled to take Jumo jet engines. After the necessary parts had been made at Rostock, the actual testing was to be done at Vienna-Schwechat, where testing of the Heinkel jet fighter was concentrated. At this point it was already clear that the ground clearance of the Junkers engines could scarcely be called more than satisfactory. In addition, the new turbines were longer than the He S8A, which would decrease the landing angle when touching down. To compensate for this, a higher landing speed had to be accepted. The power increase to 800 kp per engine, though, could not be attained otherwise.

At the beginning of March 1943 the RLM decided the time was right to make a decision for or against the He 280. On March 9 the Ministry spoke against series production of the He 280. Six days later He 280 V2, powered by two Jumo engines, made its first flight. The plane, which weighed 4800 kilograms, took off without problems. Only as it came in to land did the landing gear fail to extend completely, as the hydraulic system had been installed incorrectly. The plane came in at a good 220 kph, and despite strong braking, it shot across the landing field and onto a nearby pasture, coming to a stop there with a damaged nose wheel and bent engine cowling. After just a few days, testing could be resumed as planned. As early as April 4, 1943, test pilot Schäfer had already attained a speed of 775 kph. Despite this success, the news reached Heinkel in March that only a few models for power-plant testing would be allowed to be built. The next day Ernst Heinkel halted preparations for series production of the jet plane. The design team was transferred to more purposeful work on the He 177, so that work on the heavy bomber could finally proceed as planned. Yet the number of He 280 test models was now set at 20 planes, meaning that in addition to the seven prototypes, thirteen more were to be built. This took place just a few days before GFM Milch informed Dr. Heinkel that the He 280 was formally canceled. Frantic attempts were made at Schwechat to ready the Heinkel jet fighter for series production. Measures to eliminate the elevator vibrations in He 280 V2 and V3 were to begin with the greatest urgency. In addition, He 280 V7 was to be used to determine what flying characteristics existed at high speeds.

This photo shows the camera installed for pitch testing.

In-flight pictures of He 280 V7 during pitch testing at Ainring in October of 1944. The oversize DVL angled intake duct and the brackets for the camera can be seen clearly.

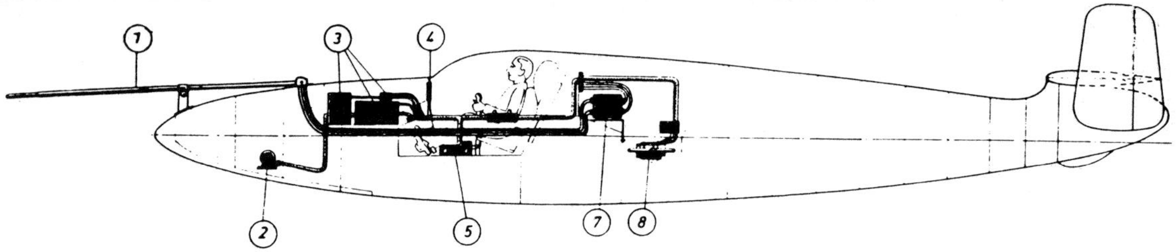

Components installed in He 280 V7 for polar measurements.

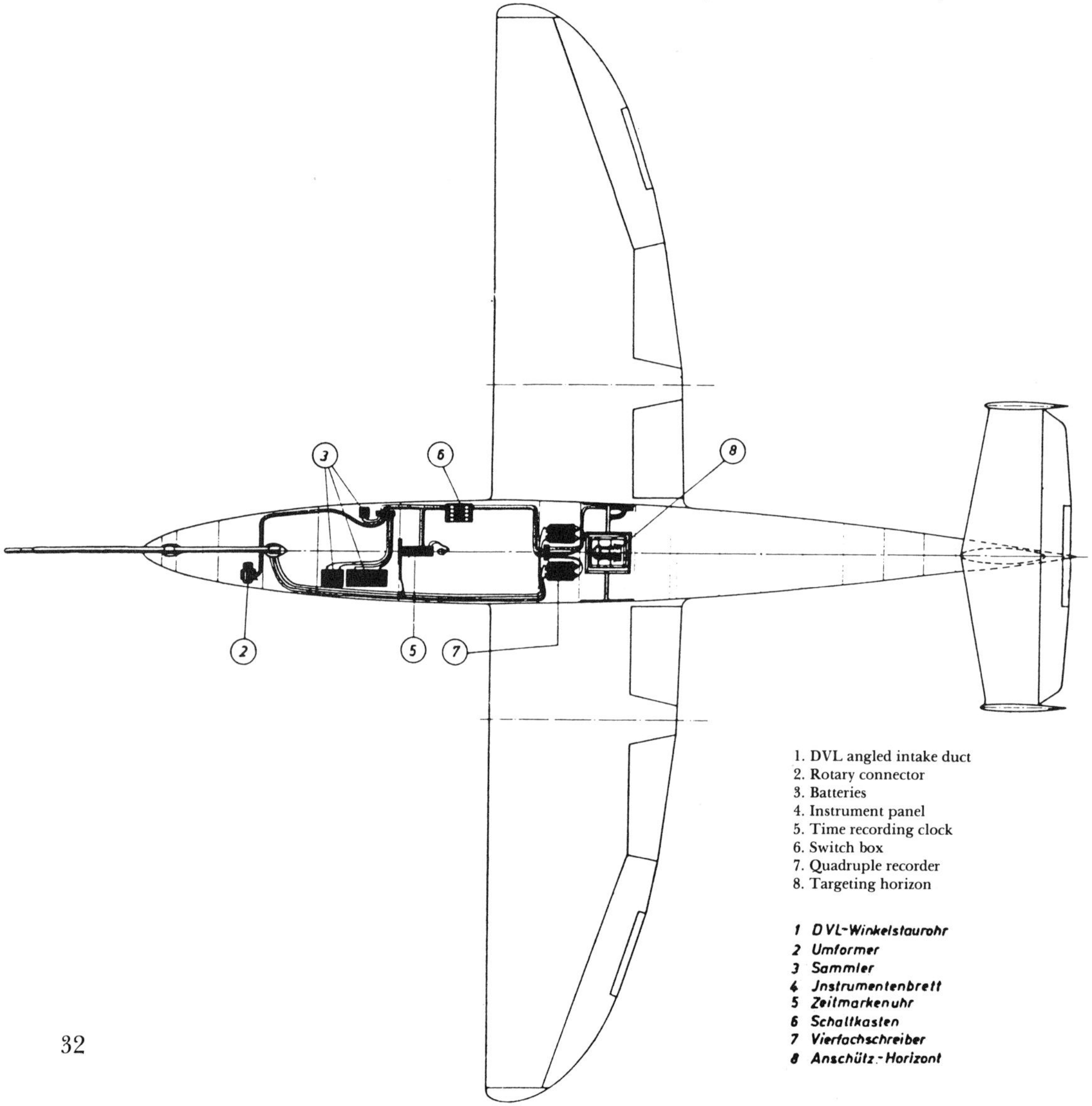

1. DVL angled intake duct
2. Rotary connector
3. Batteries
4. Instrument panel
5. Time recording clock
6. Switch box
7. Quadruple recorder
8. Targeting horizon

Above: He 280 V7 was towed into the air by an He 111.

Left: Installation drawings for polar measurements in He 280 V7. The components are identical to those used in the first test, except that there are no cameras. The tests were likewise made at Ainring in October of 1944 by the German Research Institute of Gliding "Ernst Udet."

Side views of V7 with civilian registration.

On March 27, 1943 an ominous letter arrived from the RLM, stating that development of the He 280 had been dropped in favor of the Me 262. The number of test models was immediately reduced to nine, almost all of which were to be used for powerplant testing. In addition, the decision was made to transfer personnel with vital experience in jet plane construction to Messerschmitt, so as to develop the Me 262 more quickly. Whether this actually took place is unknown.

One of the Heinkel factory buildings before the war.

A Heinkel factory memo of April 3, 1943 lists the status and planned uses of the eight He 280 planes:
— He 280 V2: Testing Jumo jet engines (in process),
— He 280 V3: Testing Heinkel jet engines (in process),
— He 280 V4: Testing Argus jets (planned for August 1943),
— He 280 V5: Testing Heinkel jet engines (planned for July 1943),
— He 280 V6: Testing BMW jet engine (planned for May 1943),
— He 280 V7: High speed test model (without engines),
— He 280 V8: Testing Jumo jet engines (planned for May 1943),
— He 280 V9: Testing BMW jet engines (planned for June 1943).

It might be noted that He 280 V4 was planned as a replacement for the crashed V1, whose schedule had to be met at all costs. Three months later, on July 6, 1943, the first two BMW engines were delivered for installation, and the scheduled date was kept. According to Schäfer, the plane was not finished in 1943.

For a scheduled demonstration at Hörsching, the cell of He 280 V4, though still unfinished, was used for part of the DFS testing. Since He 280 V8 was there at the same time, it was impossible to change the name of V4 to V8.

The first flight of He 280 V7, towed by an He 111 H of the German Institute of Gliding, took place on April 19, 1943. At an altitude of 1800 meters, Schuck attained a speed of 500 kph. The second and third flights were made without incident on the next day. The pilot of the towplane was Rudolf Opitz, while Mr. Ziegler flew He 280 V7. As of July 1943 the plane was flown by the Institute of Gliding for the purpose of polar orientation, while pitch tests and several straight-line flights completed the test program. As of September 29, 1944, comparison data for the future flight testing of He 280 V8 were to be gathered. But since the required powerplant was not available at the right time, the comparison flights had to be postponed from one month to another.

For a short time at the end of April 1943 it looked as if the He 280 still had a chance.

At that time there were, at most, three or four Me 262 prototypes ready to fly, and the first nose-wheel plane, unlike the He 280, was only planned for the autumn of 1943. Meanwhile von Ohain's team had improved the He S8's thrust considerably, to 780 kp, and prepared everything for the alternative use of the BMW P 3302. Unfortunately, BMW was not yet ready to guarantee a lasting thrust. Ten days before the first flight of Me 262 V4, on May 5, 1943, another accident happened to He 280 V2. The right

Above: The first and final assembly hall. Despite the great destruction, production continued.

Right: The paint shop. In the foreground are already sprayed, damaged Ju 88 wings. Note the unusual camouflage pattern.

engine of the plane, flown by Koenitzer, stalled. Turbine vanes were broken, long flames shot out of the engine, and at the same time the Jumo 004 began to vibrate strongly. At a speed of only 200 kph and an altitude of 15 meters, Koenitzer made a belly landing in which the bottom of the fuselage and the two engine cowlings were slightly damaged. Repairs were finished by May 21. The plane was grounded shortly thereafter for installation of an improved nose wheel and a new engine.

Unfortunately, they had no luck with He 280 V2. The Jumo engine, not yet fully developed, was to blame for the loss of the second test model on June 26, 1943. In an engine fire, the plane was 80% destroyed,

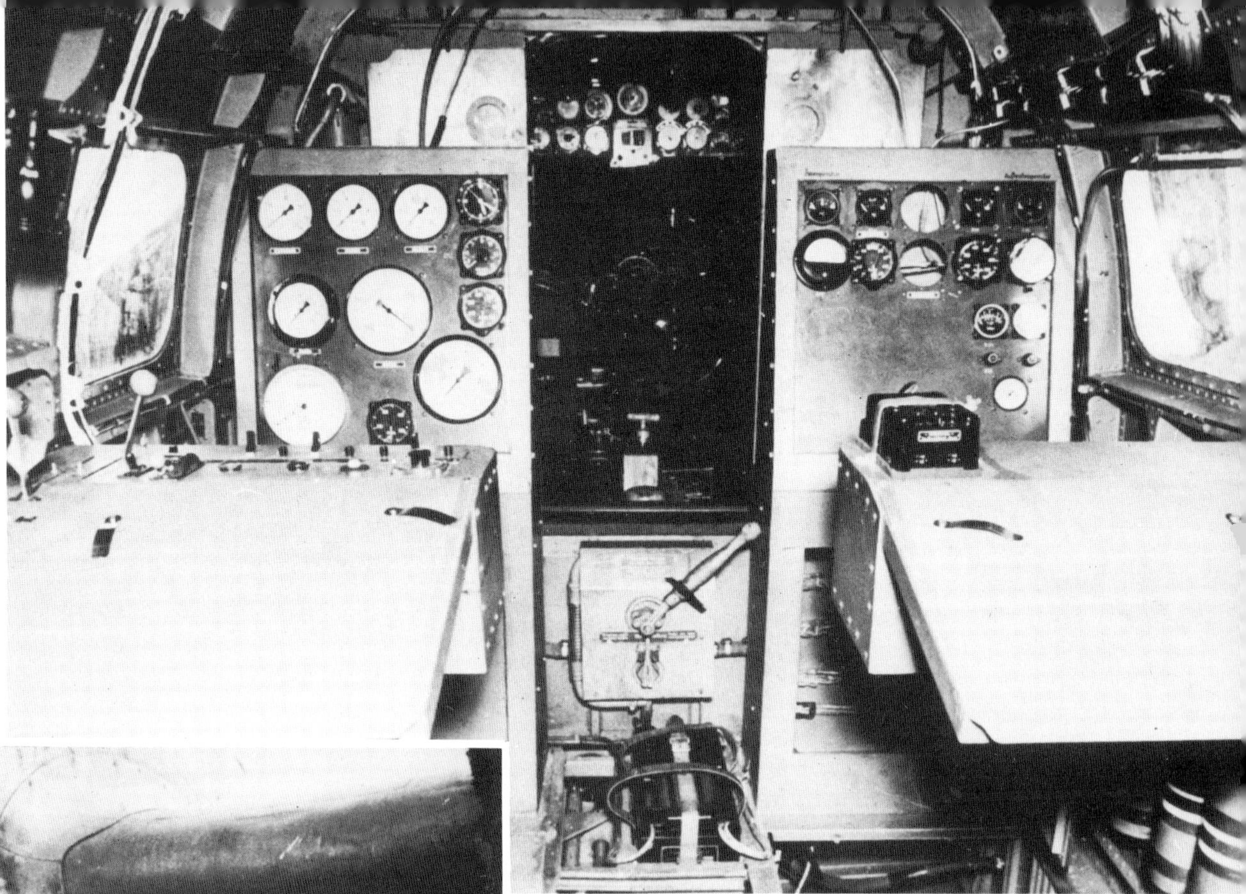

Above: Engine indicator gauges for testing the He S8A turbine engine in an He 111.

Left: Testing the He S8A engine under an He 111.

and there was no hope of rebuilding it. Thus Heinkel could fall back on only the third and seventh He 280. The V5, 6 and 9 test models were going nowhere, as the promised BMW turbine engines were not ready. The use of Argus jets, as is known, had been banned some time before. There were also new delays with He 280 V8, since a major part of the developmental capacity was needed for the He 219 night fighter. Only on July 19, 1943 did the first flight of He 280 V8, equipped with Jumo turbines, take place. By September 10, 1943, in a total of ten flights, speeds up to 750 kph were attained. The plane then had to be grounded until September of 1944, presumably for lack of replacement engines. After that it was planned to remodel it to take the V empennage, as in Heinkel's P 1073 project for a jet fighter, a short description of which was completed on July 10, 1944. In this respect it is interesting to note that Ernst Heinkel had already worked out the full details for the "People's Fighter" before the Air Ministry asked for one. In the remaining documents at the Heinkel firm, He 280 V8 does not appear again until October 5, 1944. The plane was inspected by the Technical Service at the German Institute of Gliding. After that the plane was considerably dismantled; nearly all the instruments and installations were missing.

Above: A demonstration of BMW 003 attachment under the wing.

Right: A BMW 003A jet engine, such as He 280 V4, V5 and V6 were to receive.

The pictures clearly show, from behind, the different sizes of the two engines. The Jumo 004 proved to be too big for installation in He 280. Various extensive rebuilding measures would have been required for the Jumo 004.

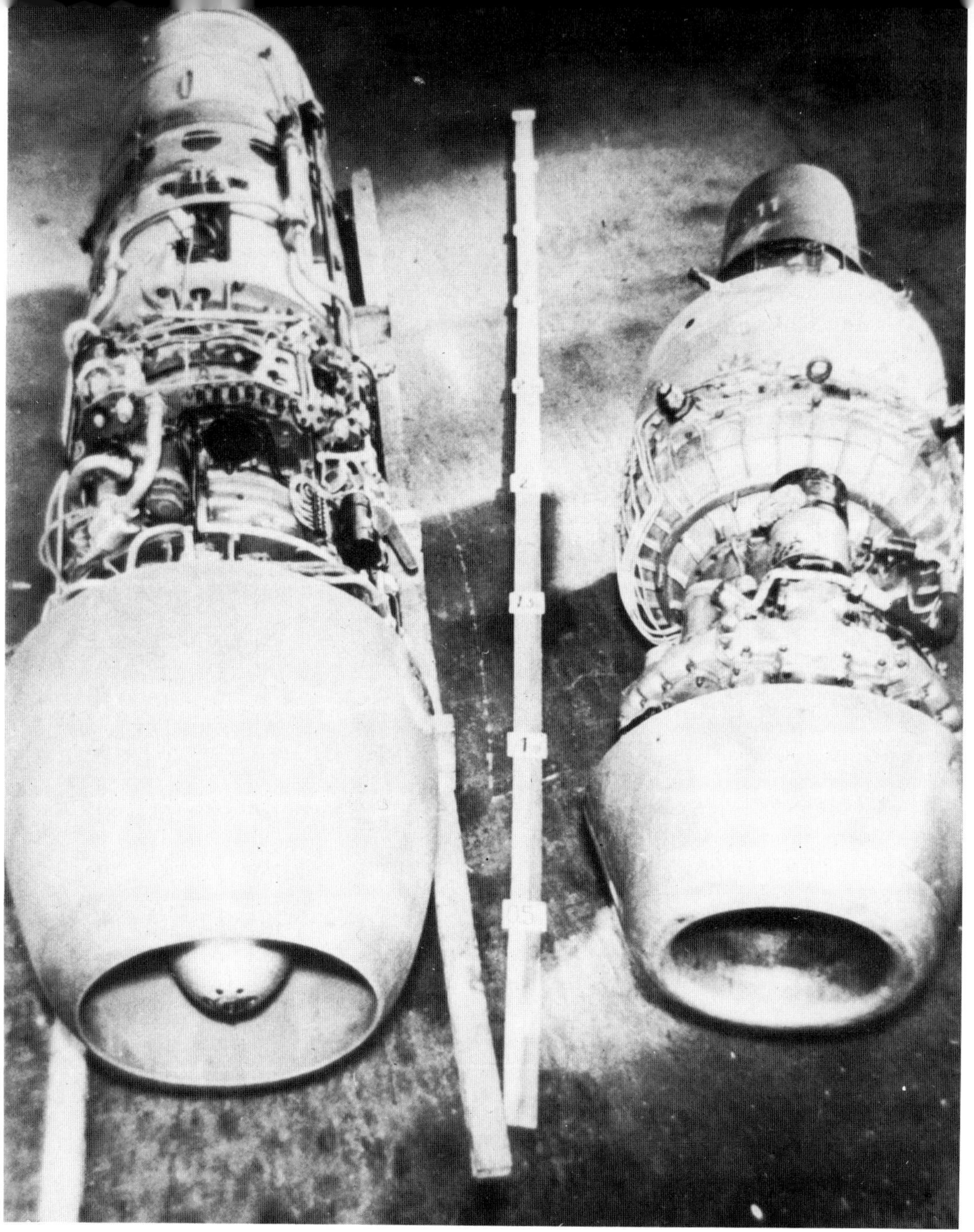

The different sizes of the Jumo 004 (left) and He S8A (right) powerplants.

Above: Side view of the He S8A powerplant.

Right: the more powerful successor to the S8A. the He S11A.

Repair work and installation of the necessary measuring equipment thus had priority in the following weeks, as the new V empennage was awaited practically any day. Tests were undertaken with a Bf 109G (factory number 14003, registration VJ+WC), as were exhaustive wind-tunnel tests. On November 29, 1944, He 280 V8 was all ready except for the still lacking engine.

So much for a look at the further history of He 280 V8; now back to 1943:

By September 10, 1943 the testing situation had not improved substantially for the He 280. From July 19 to August 5, 1944, "powerless towed flights" were scheduled for He 280 V7 behind the Institute of Gliding's He 111 (SS+KA). On July 30, 1944 it was decided in a conference at Priem, on the Chiemsee, that the wind-tunnel studies then in progress should be followed by practical testing of the V empennage with He 280 V7 and V8, since additional Me 262's were not available for testing purposes. On September 26, 1944, installation of the gauges was in full swing. The flight tests were to take place in three to four weeks, though He 280 V4 could not be included, since it was too badly damaged.

The He S011 turbine. The pictures come from a series made for exhibition purposes.

Three days later, Heinkel and the Institute of Gliding set October 20, 1944 as the date for the final commencement of testing, as far as the He 280 V8 was concerned. On October 5, though, very little had been done, other than the arrival of He 280 V4 from Vienna at Hörsching. He 280 V8 was still in a "strongly dismantled condition", with many vital parts missing. By the middle of the month, the situation looked hopeless.

On the other hand, it was possible to begin the scheduled polar measurements with He 280 V7 at the end of October. At altitudes between 3000 and 5000 meters, He 280 was released to carry out an extended flight program while gliding. Along with the pitch tests already noted, tests were carried out under transit-flight conditions, plus landings with and without control by the pilot. On November 29, 1944, He 280 V8's fate was still unclear because of the lack of engines. On January 13, 1945 the report arrived that the engine components were finally underway to the German Institute of Gliding. But for the time being, the people there waited in vain.

Above: The He S011 engine was supposed to be attached to the He 280's wings like this.

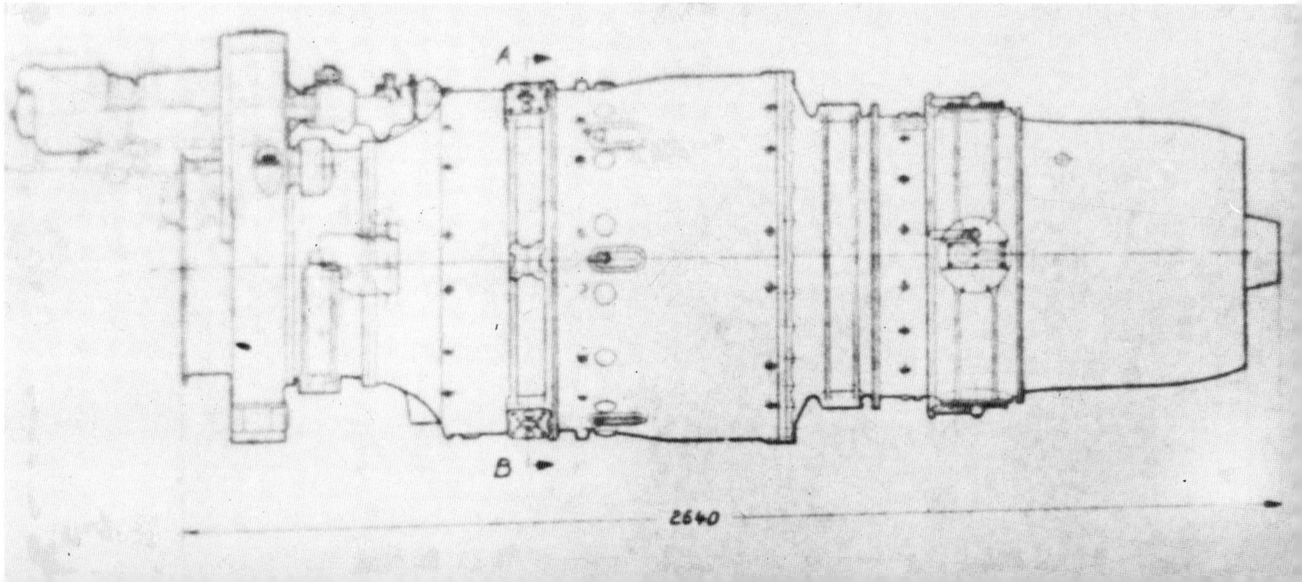

Right: An original drawing of the He S011 turbine.

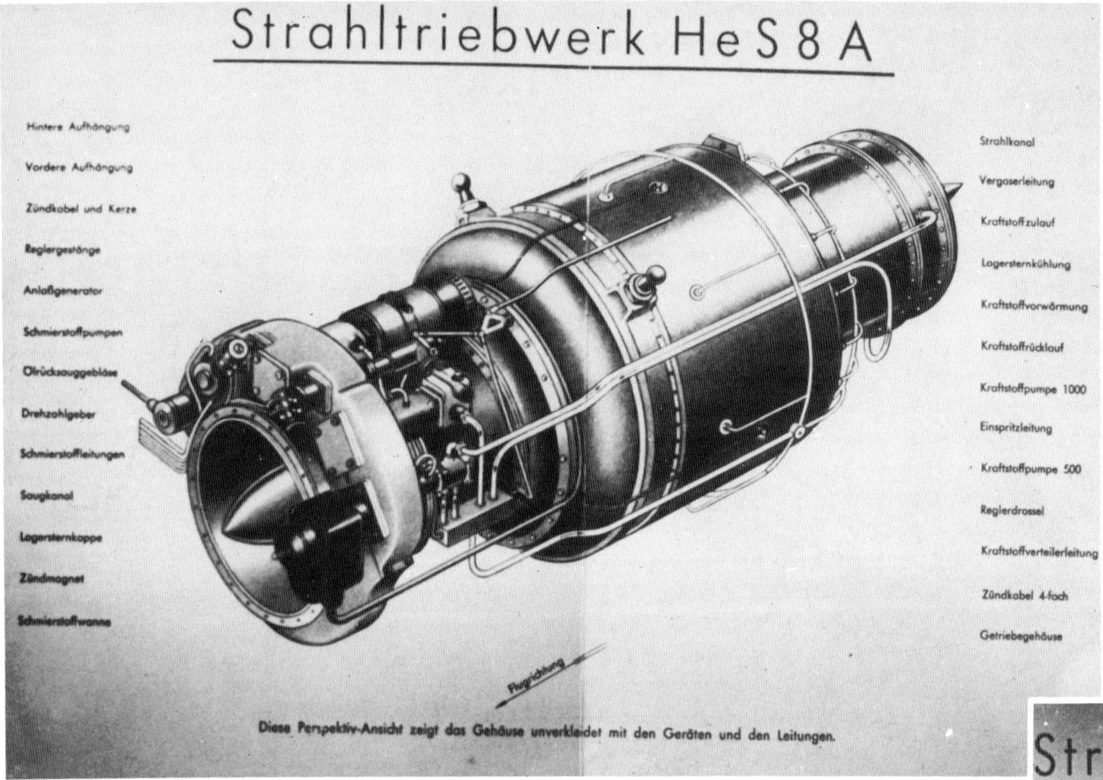

Meanwhile the fuel for the towplane ran out and was very difficult to obtain because of the war situation. Thus it was reported to the General TT (Troop Technology) in Berlin on March 14 that the plane still could not be completed, since the new V empennage "often had to be changed on account of war-related conditions", meaning that it simply could not be done. Since the He 162 with the normal jet-duct engine had satisfactory flying characteristics and sufficient stability, the work now had only an academic nature.

The He S8A engine. This is how it looked to the mechanics after the cowling was removed.

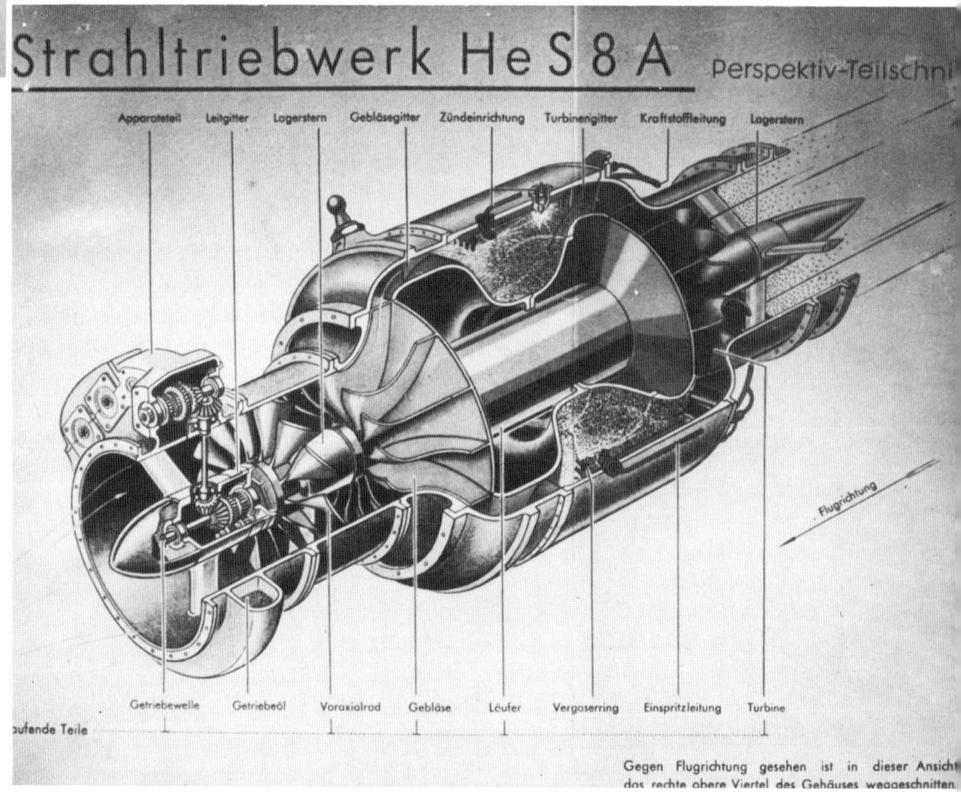

The turmoil of war had led to the loss of the first empennage during an air raid. The second had long since gone astray en route from Breslau to Hörsching. When it finally turned up, its condition was so bad that nothing could be done with it.

The Allies found the empennage of He 280 V3 at Vienna-Schwechat, and this was the last time the He 280 was mentioned. To date nothing is known of the fate of He 280 V7 or the other prototypes. Although it has been stated repeatedly that there were only three prototypes, the following is known: In October of 1944, the V4, 7 and 8 test craft were available at the same time, though

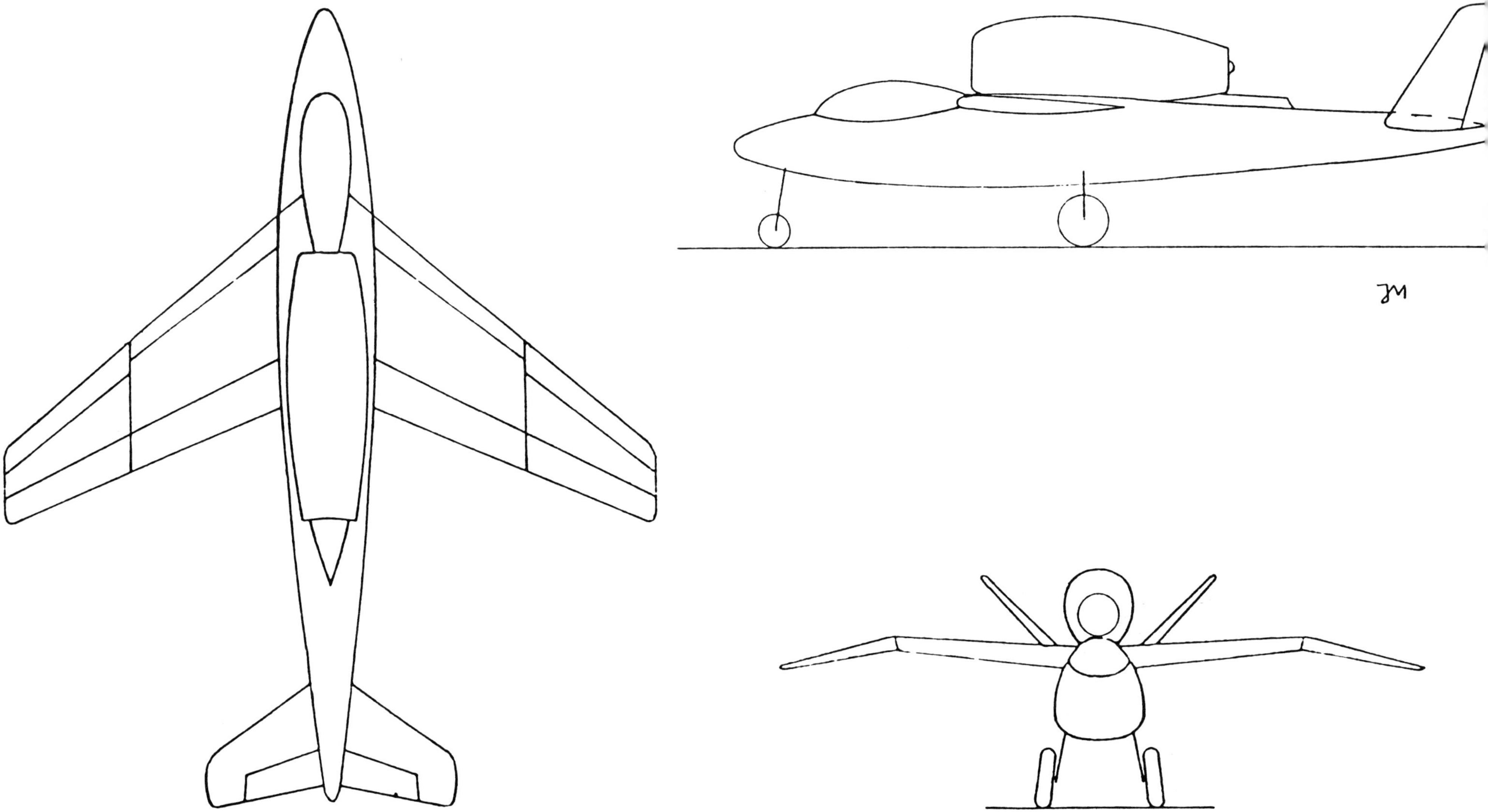

The He 162 with He S11 and V tail was another design for a Heinkel jet fighter based on the design of the He 280.

Left: Field Marshal Milch often inquired personally about the activities at the German aircraft factories and test centers.

Below: Some of Germany's greatest aeronautical pioneers visited Ernst Heinkel on July 31, 1953. From left to right: Director Rothes (Focke-Wulf), Prof.Dr. Dornier, Prof.Dr. Heinkel, Director Fiser (Dornier).

only V7 was in flying condition. The V1 and V2 prototypes had long since been scrapped after crashing. Thus there were at least five He 280's. He 280 V3, parts of which were found at Vienna-Schwechat after the war, can be added. But only parts or components of He 280 V5, 6 and 9 were manufactured. Because of lasting problems with the BMW engines, the planes were never fully assembled.

Despite all the accidents and bad luck, the He 280 was a significant step in the creation of a modern high-performance jet fighter.

HEINKEL He 280

Model	Reg.	First Flight	Engine	Notes
He 280 V1	DL+AS	9/22/1940	none	Crashed 1/13/43 after 64 towed flights
		4/1/1941 (FKT)	He S8A	
		1/13/1943	As 014	
He 280 V2	GJ+CA	3/30/1941	He S8A	80% destroyed in TL
		3/16/1943	Jumo 004	failure, 6/26/1943
He 280 V3	GJ+CB	7/5/1942	He S8A	Parts at Schwechat found May 1945
He 280 V4	–	8/31/1943 (FKT)	BMW 003	Cell at Hörsching, October 1944
		9/31/1944 (FKT)	Jumo 004	
He 280 V5	–	7/26/1943 (FKT)	He S8A	Cell not finished
		9/15/1943 (FKT)	BMW 003	
He 280 V6	–	7/26/1943(FKT)	BMW 003	Cell not finished
He 280 V7	NU+EB D-IEXM	4/19/1943	none	115 towed flights by early 1945
He 280 V8	NU+EC	7/19/1943	Jumo 004	At DFS without V tail
		8/15/1943 (FKT)	He-TL	in March 1945
He 280 V9	–	8/31/1943 (FKT)	BMW 003	Cell not finished
He 280 V10	–	4/1943 (FKT)	Jumo 004B	1st He 280 B-1 model
He 280 V11	–	5/1943 (FKT)	BMW 003	2nd He 280 B-1 model
He 280 V12	–	6/1943 (FKT)	He - TL	3rd He 280 B-1 model

(FKT: planned date on which model was to be ready to fly)

DATA SHEET He 280

		Heinkel 280 V5*	Heinkel 280 B1**
Wingspan	m	12.00	12.00
Length	m	10.40	10.40
Height	m	3.06	3.06
Wing surface	sq.m	21.50	21.50
Track	m	2.44	2.44
Engines		Heinkel-Hirth He S8V 14/15	BMW P 3302 (later BMW 003)
Sustd. thrust	kp	2 x 720	2 x 680
Armament		3 MG 151/20	3 - 5 MK 108
Gross weight	kg	3.215	3.784
Crew weight	kg	100	100
Ammunition	kg	185	106 (3 guns)
Fuel weight	kg	810	1030
Take-off wgt.	kg	4300	5020
Top speed	kph	730 near ground 770 at 6000 meters	760 near ground 815 at 6000 meters
Climbing speed	m/sec	19.1 near ground 9.6 at 6000 meters	15.9 near ground 10.7 at 6000 meters
Climbing time to 6000 m	min	8.3	7.5
Take-off	m	750	750
Service ceiling	m	11,500	13,000

* Calculated data as of July 1942
** Calculated data as of December 1942

Left page: The test results of the He 280 were applied directly to the production of the He 162.

Schiffer Military History

Schiffer Publishing Ltd.
1469 Morstein Rd.
West Chester, PA 19380
Phone: (215) 696-1001 FAX: (215) 344-9765

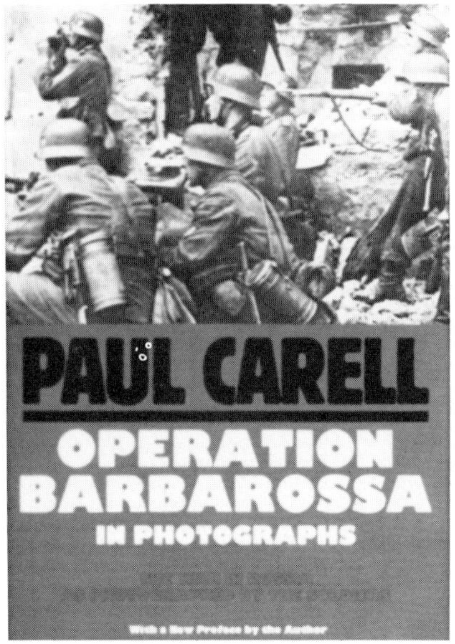